GROSS STUFF
that Helps Ocean Animals Survive

Lee Ann Landstrom AND Karen I. Shragg
ILLUSTRATED BY Rachel Rogge

Mountain Press Publishing Company
Missoula, Montana
2023

AUTHORS' NOTE

We chose the sea as the focus of this fourth *Nature's Yucky!* book to highlight the problems of the oceans on our watery planet. As yucky as some of the behaviors of sea animals can be, they are an interconnected part of a vast underwater ecosystem we are just beginning to understand. We enjoy the "yucky" learning format, developed in our first three *Nature's Yucky!* books, to introduce wildlife to children. As readers will learn, even people living inland have huge impacts on the vast oceans—because of the trash we throw out, the packaging we choose, the energy we consume, and the fish and other seafood we eat. We hope our young readers and their families will become interested in multiple conservation efforts to help save our oceans and the wonderful species in them.

Library of Congress Cataloging-in-Publication Data

Names: Shragg, Karen, 1954- author. | Landstrom, Lee Ann, 1954- author. | Rogge, Rachel, 1971- illustrator.
Title: Nature's yucky! in the sea : gross stuff that helps ocean animals survive / Karen I. Shragg and Lee Ann Landstrom ; illustrated by Rachel Rogge.
Missoula, Montana : Mountain Press Publishing Company, 2023. | Includes bibliographical references. | Audience: Ages 5 and up |
Identifiers: LCCN 2023007529 | ISBN 9780878427109 (paperback)
Subjects: LCSH: Marine animals—Miscellanea—Juvenile literature. | Seashore animals—Miscellanea—Juvenile literature. | Marine animals—Behavior—Juvenile literature. | Seashore animals—Behavior—Juvenile literature.
Classification: LCC QL122.2 .S45 2023 | DDC 591.77—dc23/eng/20230330
LC record available at https://lccn.loc.gov/2023007529

PRINTED IN HONG KONG BY MANTEC PRODUCTION COMPANY

P.O. Box 2399 • Missoula, MT 59806 • 406-728-1900
800-234-5308 • info@mtnpress.com
www.mountain-press.com

To my beloved late mother, Sarah Bernstein Shragg,
for always being such a great cheerleader. —KS

To my late mother, Shirlee I. Landstrom, 94 years young,
who was always so proud of her "yucky" daughter. —LAL

For Patience. —RR

Did you know. . .

that Spot-fin Porcupinefish,

those bloated, blimp-like, beguiling fish,

fill their stomachs with water to swell up

and become more difficult to swallow?

But hey, it's okay.
Just imagine if it weren't that way!

If it weren't that way, these prickly puffers would be eaten by hungry prowlers. When threatened, porcupinefish fill their stomachs with water and stick out their spines. Predators may choose a smaller, less-spiny meal. This defense system works well against all predators except those with large mouths, like sharks. Having an unappetizing look, or being too big and too prickly, is a great way to live to see another day.

But perhaps not the best idea for a Halloween costume!

Did you know. . .

that California Moray Eels,

those cunning, coral-dwelling, carnivorous fish,

bite off the legs of crabs while they are still alive?

But hey, it's okay.
Just imagine if it weren't that way!

If it weren't that way, moray eels would not have enough food to survive. Moray eels hang out in the crevices of coral reefs, where their camouflage colors help them stay hidden. The hungry eels are ready to strike at any unsuspecting prey. If an unlucky crab wanders by the hideout of one of these sharp-toothed critters, the crab will get its legs bitten off because that is likely the closest thing the eel can reach. The crab may be too big to eat all at once, so it must be eaten one leg at a time. Aren't you glad you aren't on an eel's menu?

Did you know. . .

that Sea Lampreys,

those impossibly invasive, injuring parasites,

gnaw holes in the flesh of living fish?

Eeewww!! That's

But hey, it's okay.
Just imagine if it weren't that way!

If it weren't that way, sea lampreys wouldn't survive in the Atlantic Ocean, where they normally live. These unique, eel-like fish have cartilage instead of bones and a sucking, teeth-lined disk for a mouth. This strange-looking creature hasn't changed much for 340 million years, so its strategy of gnawing the flesh of live fish must be working. In the ocean they do not kill the fish they latch onto, but when they get into freshwater bodies, like the Great Lakes, lampreys can kill lake trout and other freshwater game-fish. The yuckiest part about these invasive species is they can wipe out local fish populations. They truly are what fish nightmares are all about.

Did you know. . .

that **Lightning Whelks**,

those curvy, conniving clam killers,

drill holes into clam shells and turn

clam guts into a soup they can eat?

But hey, it's okay.
Just imagine if it weren't that way!

If it weren't that way, these innocent-looking sea snails would not be able to eat. We often think of seashells as just something pretty to collect, but they are really animals. Lightning whelks eat other shelled creatures by drilling a hole with their sharp tongue. Then, they inject a chemical that turns the insides of clams or other hard-shelled animals into a liquid, so that whelks can drink the guts like you might drink a shake with a straw. The shells of dead whelks serve as homes for hermit crabs, so it's important to leave them on the beach. Next time you see a beautiful shell on the beach, just take a close look or a photograph and keep on walking.

Did you know. . .

that **Pacific Herring**,

those noisy, nonchalant navigators,

make fart noises to communicate?

But hey, it's okay.
Just imagine if it weren't that way!

If it weren't that way, herring could not effectively communicate with the thousands of other herring that move with them in a group called a school. Scientists discovered that these small fish take a gulp of air at the surface of the ocean and force it out of their butts. Herring make this farting noise only at night. Their high-pitched farts keep them safer because they can communicate with each other and cannot be heard by other fish, although whales and dolphins can hear them. Some fish communicate with noises from an organ called a swim bladder located on their sides, but only herring seem to talk by farting. You'll be sent to your room if you try speaking this way at the dinner table.

Did you know. . .

that Sharknose Gobies,

those puny, parasite-eating predators,

munch on nasty critters that live in the gills of fish?

Eeewww!! That's Yucky!

But hey, it's okay.
Just imagine if it weren't that way!

If it weren't that way, the coral reef and its fish populations would not be as healthy or clean. Gobies eat dead tissue and small animals called parasites that live on larger fish, leaving the fish much healthier. Because both the tiny goby and the larger fish are better off, scientists call this relationship mutualism. Scientific studies show there are more fish, more kinds of fish, and healthier fish when these little cleaning machines are present. Even in aquariums, fish will line up to be cleaned by one of these yucky but amazing gobies as if they were going to a car wash. Aren't you glad people invented washcloths?

Did you know. . .

that **Sand Tiger Sharks**,

those cannibalistic, cagey carnivores,

eat their baby brothers and sisters

while still inside their mother?

But hey, it's okay. Just imagine if it weren't that way!

If it weren't that way, baby sand tiger sharks wouldn't be able to get big enough to survive predators once they are born. Unlike some other shark species, these sand sharks give birth to live young called pups. The ones who hatch first inside the mother will become the biggest by eating smaller embryos and eggs. Once they are born, they are already big enough to be off the menu of other hungry fish. This eating of sibling embryos assures that sand tiger sharks can continue to live as part of the ocean community. Gives a new meaning to sibling rivalry, doesn't it?

Did you know. . .

that **Black Swallowers**,

those gorging, gluttonous, gassy fish,

may eat themselves to death?

Eeewww!! That's

But hey, it's okay. Just imagine if it weren't that way!

If it weren't that way, they may not get enough food in the deep, open ocean, where food is harder to find. Black swallowers eat whatever meal comes their way, even though this may be risky. Their jaws swing downward, allowing them to eat prey bigger than their heads. Their stomachs stretch out twice their body length and ten times as big! If swallowers don't digest all the food before it starts to decompose (rot), gases form. They die because their bloated stomachs may explode, and the lightness of the gas forces them to the ocean surface like an inflated beach ball. After a big Thanksgiving meal, you may feel bloated like a black swallower, but don't worry, you won't explode!

Did you know. . .

that **Caribbean Reef Squids,**

those color-shifting, surprising, shiny swimmers,

squirt a cloud of nasty black ink

at their predators?

But hey, it's okay.
Just imagine if it weren't that way!

If it weren't that way, these boneless, shell-less mollusks could be caught and eaten more often. Squids hang out in schools of four to thirty individuals, so there are many eyes to watch for predators. But if a large fish (or a human snorkeler) gets too close, squids squirt out a dark, inky liquid that confuses the predator and masks the squids' rapid getaway. Though it may sound cool, don't even think of squirting ink at home.

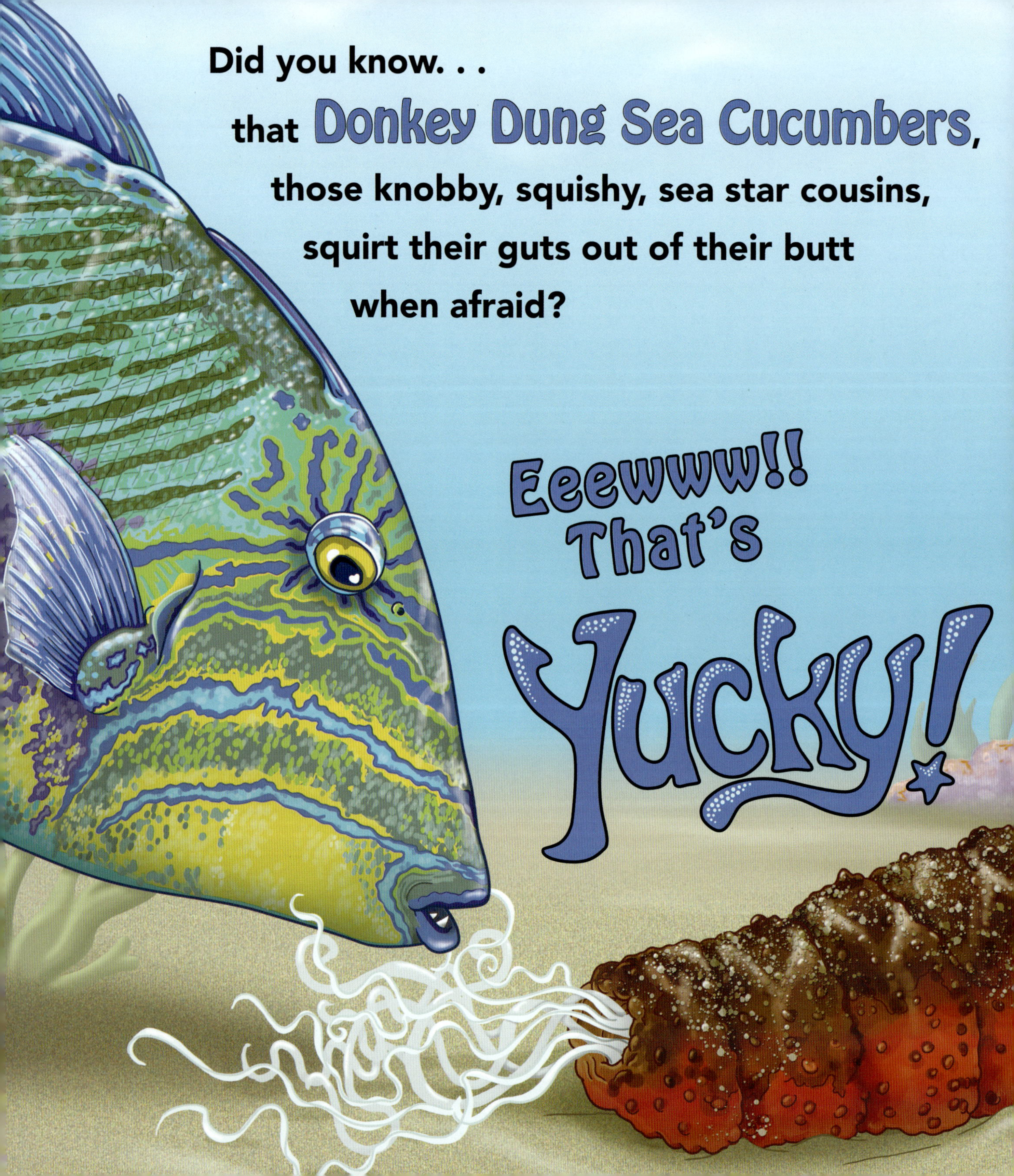
Did you know. . .
that Donkey Dung Sea Cucumbers,
those knobby, squishy, sea star cousins,
squirt their guts out of their butt
when afraid?
Eeewww!! That's
Yucky!

But hey, it's okay.
Just imagine if it weren't that way!

If it weren't that way, they would be easier prey for their enemies. These slow-moving sluglike inhabitants of the seafloor use this distraction to get away. If annoyed, squeezed, or threatened, sea cucumbers violently contract their body muscles and shoot out dozens of string-like tubules that entangle their attackers. Then sea cucumbers crawl away on their many "tube" feet to burrow or hide. The missing tubules grow back in a week or two. Gives new meaning to "spilling your guts," doesn't it?

Did you know. . .

that **Atlantic Pearlfish**,

those super-adapted,

sneaky, squatters,

live inside the Donkey Dung

Sea Cucumber's butt?

But hey, it's okay.
Just imagine if it weren't that way!

If it weren't that way, these slender fish wouldn't have a safe home. Baby pearlfish, called larvae, are free-swimming for three months before they find their sea cucumber host and move inside. Once within, young pearlfish never leave the sea cucumber; they survive by eating the sea cucumber's innards, which fortunately grow back quickly. When the Atlantic pearlfish are grown up, they leave the sea cucumber every night to look for food. Then they return to their strange mobile home in the morning. At least they don't hang pictures inside!

Did you know. . .

that **Yellowline Arrow Crabs**,

those spider-like, spindly-legged crustaceans,

eat their prey alive, one little bite at a time?

But hey, it's okay.
Just imagine if it weren't that way!

If it weren't that way, these stick-like crabs with teeny pinchers and tiny mouths couldn't eat their larger prey. These spindly crabs grab and hold marine worms with their long, spiny legs. Their razor-sharp, tiny pincers pick away, snipping the food into bite-sized pieces. Arrow crabs give being a picky eater a whole new meaning.

Did you know. . .

that California Halibuts,

those silly looking,

sideways-swimming sand dwellers,

have one eye that

moves to the other side of their heads?

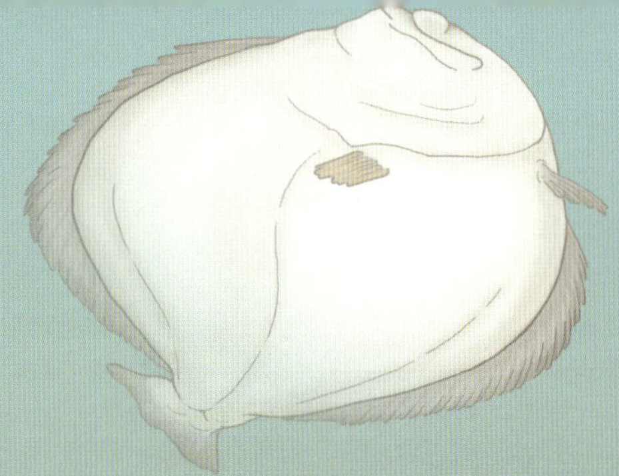

But hey, it's okay.
Just imagine if it weren't that way!

If it weren't that way, these bottom-dwelling flatfish couldn't see as well above them. Newly hatched halibuts, like their sole and flounder cousins, look and swim like normal fish, with one eyeball on each side of their heads. Then something amazing happens! When they are twenty days old and one inch long, one eye starts moving to the other side. Soon these fish begin living on the bottom of the ocean, lying on their eyeless side. They hide in the muck with both eyes looking upward, ready to snatch smaller fish for dinner. Imagine trying to fit them with eyeglasses!

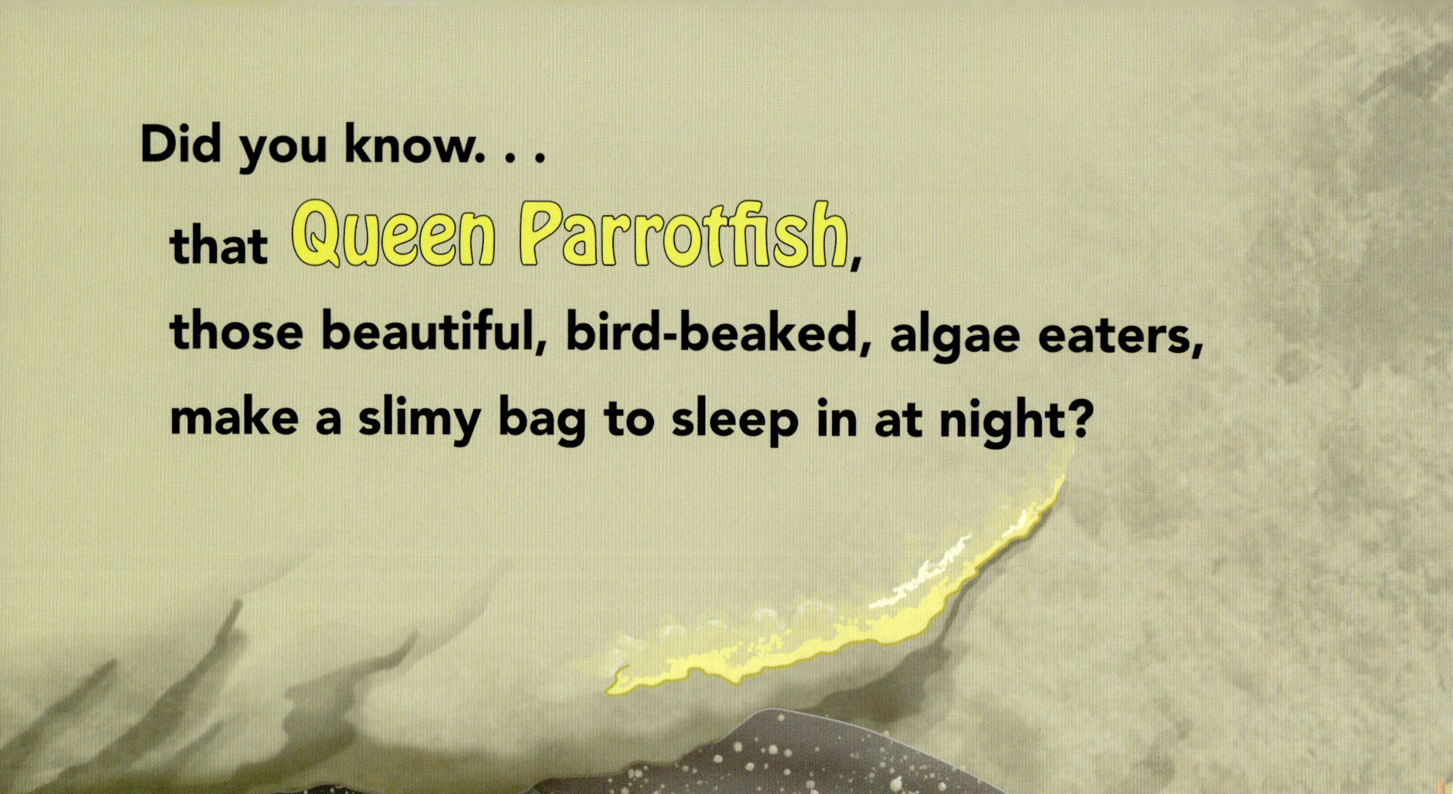

Did you know. . .

that **Queen Parrotfish**,

those beautiful, bird-beaked, algae eaters,

make a slimy bag to sleep in at night?

But hey, it's okay.
Just imagine if it weren't that way!

If it weren't that way, the sleeping parrotfish would be easier to detect and catch by their enemies out hunting them at night. All species and sizes of parrotfish find a nook in rocks or coral and wedge themselves in at night for safety. Some species, like the Queen, also secrete a mucus bubble to cover their bodies in a slimy cocoon. The mucus disguises the brightly colored fish from eels and sharks, and it also hides their scent. Aren't you glad that sleeping bags aren't made of slime?

Did you know. . .

that **Stonefish**,

those secretive, super-disguised sitters,

can badly hurt a person with their

poison-tipped spines?

But hey, it's okay.
Just imagine if it weren't that way!

If it weren't that way, these rock imitators couldn't survive as well. The highly poisonous back spines help protect them from predators. Stonefish look like coral or muddy rubble with their bumpy shapes, skin flaps, and blotches of color. People swimming or wading may fail to see the motionless, camouflaged stonefish, and if they touch or step on the spines, they get a very painful, sometimes deadly sting. Beware: Not all rocks are really rocks!

that **Pacific Hagfish**,
those snake-shaped, freaky-looking fish,
produce gallons of slime to get away
from predators?

But hey, it's okay.
Just imagine if it weren't that way!

If it weren't that way, they couldn't escape their enemies. If attacked or stressed, these jawless, blind fish can produce five gallons of slime in five seconds. This mucus tastes bad and is so thick that it jams up the mouth and gills of the attacker, who then lets go and swims away to clear its gills. Imagine the trouble you could get into with all that slime!

Fun Facts about Animals of the Sea and Ocean

SPOT-FIN PORCUPINEFISH *Diodon hystrix*

Size: 16 to 36 inches (40.6 to 91.4 centimeters)

Food: crustaceans and mollusks, such as snails, crabs, sea urchins, or worms

Range: coral reefs, lagoons, and sandy or muddy bottoms in temperate and tropical seas worldwide; typically found 10 to 65.5 feet deep (3 to 20 meters), down to 164 feet (50 meters)

Spot-fins, also known as black-spotted or spotted porcupinefish, are nicknamed blowfish, balloonfish, or globefish because they puff out their bodies and spines. The porcupinefish swallow extra water, sometimes doubling their size, so they might be too big for the predator to eat. When inflated, they look like a porcupine and quite scary to any fish that would dare eat it. These solitary, nocturnal fish hide in caves, holes, or under ledges and coral in the day. Porcupinefish are considered a delicacy to eat in Japan and the Philippines, but great care must be taken during preparation because the fish has poisonous skin and organs. These unusual fish are found worldwide because their eggs are laid in the open ocean and travel on currents for five days before they hatch.

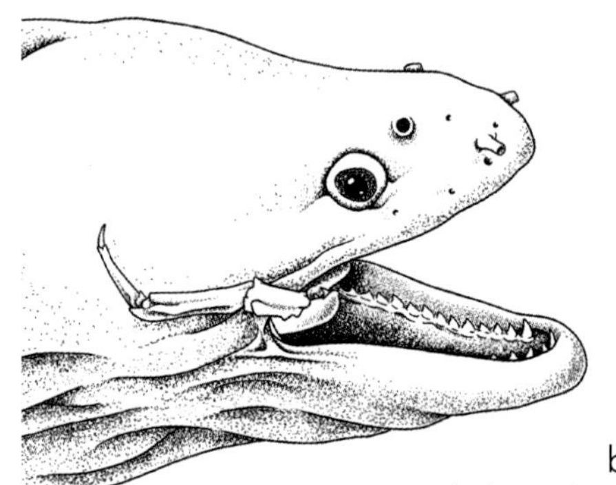

CALIFORNIA MORAY EEL *Gymnothorax mordax*

Size: 5 feet (1.5 meters)

Food: small reef fish, crabs, squid, shrimp, octopi, and sea urchins

Range: rocky areas of the eastern Pacific coast from Point Conception in southern California to the Baja peninsula in Mexico; surface down to 131 feet (40 meters)

All eels have two sets of jaws full of sharp teeth: one set you can see, the other is located deeper in its throat and helps the eel to eat its prey in smaller bites. Morays look like they are gasping because they must constantly open and close their mouths to force water over their gills so they can breathe. Eels can live to be thirty years old. Morays secrete a yellowish mucus for protection against rough rocks. The largest morays weigh 66 pounds (30 kilograms).

SEA LAMPREY *Petromyzon marinus*

Size: 23.5 to 29.5 inches (60 to 75 centimeters), up to 47 inches (120 centimeters)

Food: blood, body fluids, and meat of many fish species

Range: native in the north Atlantic Ocean, the Mediterranean Sea, parts of New York state, Vermont, and Connecticut; invasive in the Great Lakes and inland streams; found down to 2.5 miles (4,000 meters)

Sea lampreys are primitive, boneless, jawless animals with disk-like mouths used to chew on their prey. Like salmon, sea lampreys migrate to freshwater to spawn and live most of their lives in the ocean. They were first found in Lake Ontario in the 1800s and had moved to the other Great Lakes by 1921. They kill native fish such as lake trout in the Great lakes, so there are control programs that have reduced lamprey populations by 90 percent. They are a food delicacy in parts of Europe.

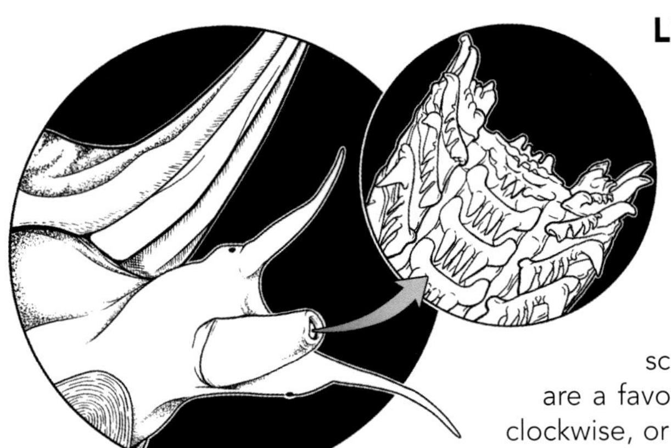

LIGHTNING WHELK *Sinistrofulgur perversum*

Size: 2.5 to 16 inches (6 to 40 centimeters)

Food: oysters, clams, and scallops

Range: shallow, sandy, or muddy bays, and seagrass beds in the Gulf of Mexico and the Atlantic, from North Carolina to Florida

Whelks have special sensors to "smell" their food. Their egg castings look like snakeskin. Lightning whelks were used by Native Americans as a food source, and their shells were used for making scrapers, bowls, and jewelry. The beautiful shells of lightning whelks are a favorite of collectors because their spiral shell twists uniquely counterclockwise, or to the left. They are the state shell of Texas. It's important to know that these animals are part of the seashore ecosystem.

PACIFIC HERRING *Clupea pallasii*

Size: 10 to 18 inches (25 to 45 centimeters)

Food: tiny organisms called phytoplankton, zooplankton, small crustaceans, and little fish

Range: coastal north Pacific (North America to northeast Asia); found down to 492 feet (150 meters)

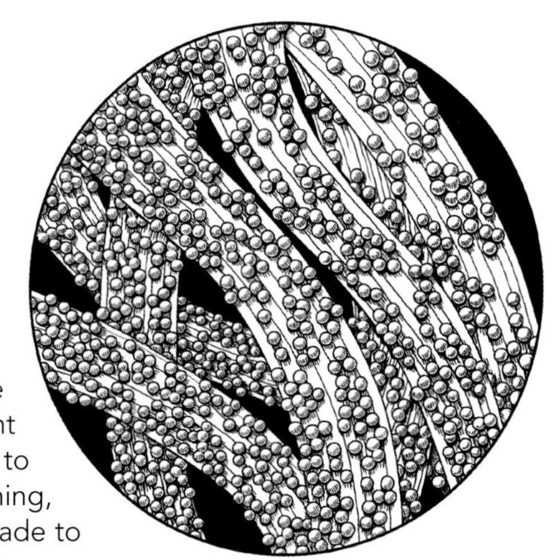

Pacific herring lay up to 20,000 eggs on eelgrass, but very few eggs reach maturity because there are so many predators. Sea mammals, birds, and larger fish depend on the abundance of this small fish and eat the eggs, babies (larvae), and adults. Herring are an important keystone species in the food chain, meaning that many other animals depend on this formerly abundant fish. Many humans also love to eat herring, and the eggs (roe) are a delicacy to millions in Asia. In 1993, the Pacific herring fishery collapsed due to overfishing, and it is slowly recovering thanks to conservation efforts. Efforts are being made to have them listed as endangered or threatened to offer them more protection.

SHARKNOSE GOBY *Elacatinus evelynae*

Size: 1.6 inches (4 centimeters)

Food: parasites on fish, copepods, and brine shrimp

Range: coral reefs in tropical western Atlantic, Gulf of Mexico, and Caribbean Sea; found 3 to 174 feet deep (1 to 53 meters)

Sharknose gobies, also called neon or Caribbean cleaning gobies, are a favorite of the aquarium trade because of their beautiful colors: yellow and blue, yellow, or yellow and white. They live on coral reefs where pairs of gobies will mate for life. The male watches over the eggs, eating those that develop a fungus. Parental duties are over as soon as the eggs hatch. As parasite eaters, these hard-working cleaners have a ready source of food—while the fish they are "cleaning" get a free parasite-removal service.

SAND TIGER SHARK *Carcharias taurus*

Size: up to 10.5 feet (3.2 meters)

Food: bony fish, crustaceans, skates (a small ray), squid, and other sharks

Range: tropical, sandy bottoms or reefs in the Atlantic and Pacific Oceans, Indian Ocean, and the Mediterranean Sea; often found 49 to 82 feet deep (15 to 25 meters), down to 630 feet (191 meters)

Sand tiger sharks are unique in the shark family because they can gulp air at the surface, which allows them to float effortlessly while hunting for food at night. They hide in coves and rock beds during the day. Although they appear menacing and are related to the great white shark, these docile sharks are quite harmless to humans. The decline in the numbers of this vulnerable species has to do with their low birth rate, people's use for their oil and skin, and the increasing demand for their fins for shark fin soup, a delicacy in Asia.

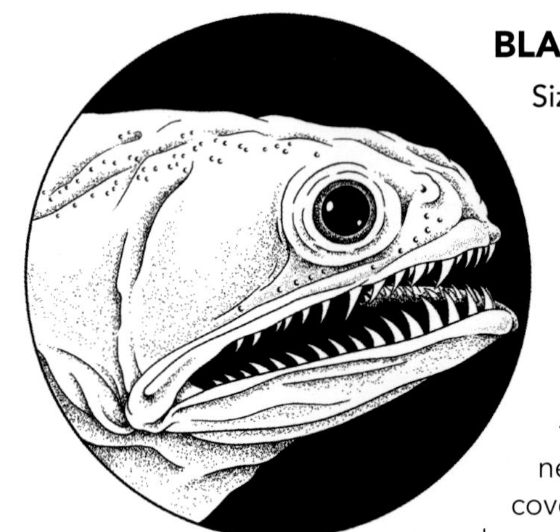

BLACK SWALLOWER *Chiasmodon niger*

Size: up to 10 inches (25 centimeters)

Food: fish, often longer than itself

Range: tropical and subtropical waters worldwide, especially in the north Atlantic; found in the deep sea: 490 to 12,795 feet (150 to 3,900 meters), usually deeper than 2,296 feet (700 meters)

These small, elongated, flattened fish have sharp, interlocking teeth and no scales. Their skin is so black the fish are nearly invisible in the deep dark sea. Their eggs float in the open ocean, and these egg masses are commonly found off South Africa. Young swallowers (juveniles) are found in the summer near distant Bermuda in the Caribbean. The baby fish (larvae) and juveniles are covered with small, projecting spines. Although common and widespread in the deep sea, they are not used for human food.

CARIBBEAN REEF SQUID *Sepioteuthis sepioidea*

Size: up to 10 inches (25 centimeters)

Food: small fish such as sardines, and shrimp

Range: in the shallows around coral reefs and seagrass flats throughout the Caribbean but not in the Gulf of Mexico; typically found 4.5 to 25 feet deep (1.3 to 7.6 meters), and as deep as 329 feet (100 meters)

Reef squids use more than forty colors, patterns (stripes, spots, blotches), and flashing to communicate with other squids in their school. They can even show a different message on each side! Curious squids watch snorkelers and may not retreat until people are only 5 feet away. When danger approaches, they turn dark or pale to confuse or distract predators, make fake large eyespots, or jet away, often squirting black dye. Squids have 8 arms and 2 longer tentacles that grab prey. They are big eaters, often eating half their body weight each day. In restaurants, squids are referred to as "calamari."

DONKEY DUNG SEA CUCUMBER *Holothuria mexicana*

Size: up to 20 inches (51 centimeters)

Food: sifting through sediments to eat buried organic stuff like algae, plankton, tiny organisms, dead animals, or waste matter

Range: shallow calm waters with sandy bottoms, seagrass beds, offshore reefs, rocky terraces, lagoons, and mangroves in the Caribbean, south to Brazil and the Azores; typically found 3 to 33 feet deep (1 to 10 meters), down to 66 feet (20 meters)

Unlike their sea star cousins, sea cucumbers have back and front ends, with sensory tentacles at the front. If a part of a sea cucumber is eaten or removed, it can grow back the missing part. These "vacuum cleaners of the seafloor" are efficient recycling machines because they eat algae and waste matter. All sea cucumber species have declined globally, mostly due to overfishing of these easily caught creatures. Costa Rica, Panama, and Venezuela have banned the catching of donkey dung sea cucumbers. Many other sea cucumber species, however, are legally and illegally sent to China as a popular so-called health food.

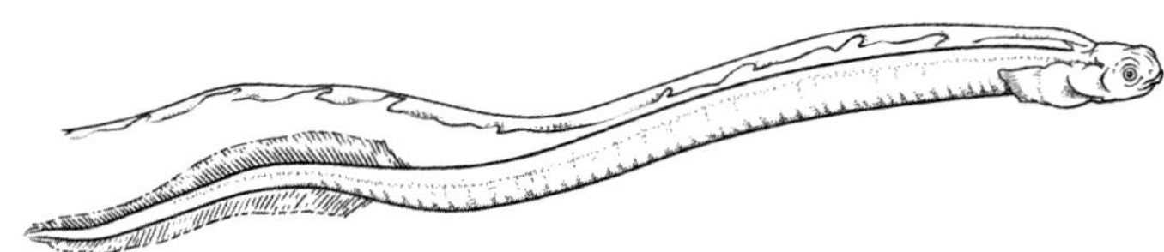

ATLANTIC PEARLFISH *Carapus bermudensis*

Size: up to 9 inches (23 centimeters)

Food: when young, they eat the insides of sea cucumbers; adults eat small shrimp, worms, isopods, crabs, other creatures, and even dead fish

Range: mostly lives inside sea cucumbers in reefs or seagrass beds in the Caribbean and to the north coast of South America; typically found 3 to 33 feet deep (1 to 10 meters), down to 770 feet (235 meters)

Most pearlfish live symbiotically, meaning they don't hurt their sea cucumber host while using it as a daytime shelter. Other species of pearlfish live inside clams, conches, starfish, or sea squirts. Several other types of pearlfish never leave the host; they only eat the sea cucumbers' insides, which fortunately regrow quickly. Although pearlfish mostly live alone, one scientist found fifteen pearlfish living inside one sea cucumber! These eel-like, nocturnal fish make pulsing sounds that can be heard for long distances to locate mates.

YELLOWLINE ARROW CRAB *Stenorhynchus seticornis*

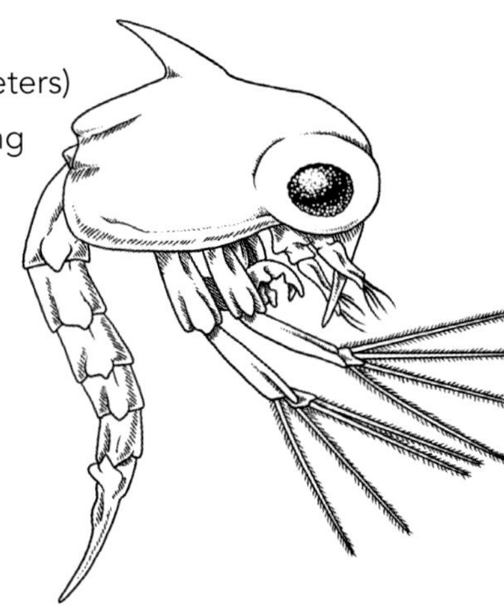

Size: shell size 1 to 2.5 inches (2.5 to 6 centimeters), with legs, up to 6 inches (15 centimeters)

Food: live feather duster worms, shrimp, and other small creatures; sometimes scavenging

Range: rocky, gravel, or sandy bottoms or coral in the Caribbean and Atlantic from North Carolina to Argentina; found 10 to 30 feet deep (3 to 9 meters)

These colorful crabs, sometimes called spider crabs, have very slender, spider-like legs, three times longer than their bodies. Like all crustaceans, they outgrow and shed their shell, which is often eaten for extra energy. If a leg is lost, they can grow a replacement leg over time. These nocturnal and territorial crabs hunt small feather duster worms and other live tiny animals, and also scavenge. In fact, arrow crabs are often kept in aquariums to eat and control pesky marine worms. These crabs are also found hiding in the long, reaching "fingers" or stinging tentacles of sea anemones during the day. The tentacles protect the small crabs from larger, hungry fish and eels.

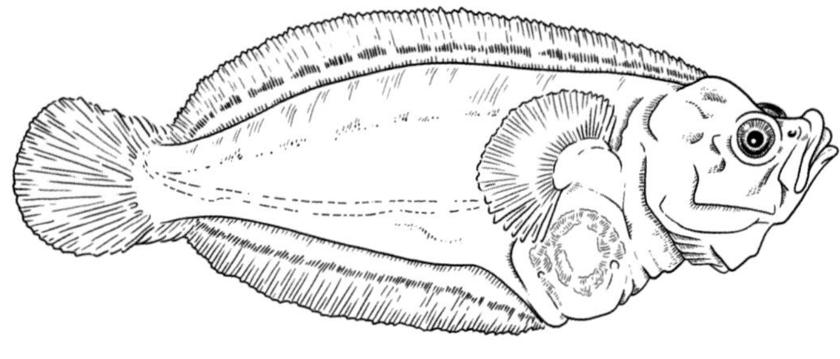

CALIFORNIA HALIBUT *Paralichthys californicus*

Size: up to 60 inches (1.5 meters)

Food: small fish, particularly sardines and anchovies, octopi, and squid

Range: soft sand or mud bottoms, sometimes kelp (seaweed) beds or rocky areas, nearshore shallows, and estuaries in the eastern Pacific from British Columbia to Baja; typically found 10 to 80 feet deep (3 to 24 meters), down to 600 feet (183 meters)

Halibut bury themselves in the sand or mud to hide from their predators and to ambush their prey, which they will chase to the surface if necessary. Like their flounder cousins, they can change color to match the seafloor. Unlike their flatfish relatives whose second eye moves to one specific side, California halibut are 40 percent "left-eyed" and 60 percent "right-eyed." As adults, they swim horizontally— not upright—in a fluttering, waving motion. Small halibut are nicknamed "flyswatters," but over their thirty-year lifespan, they can grow as big as barn doors. Halibuts are a popular but regulated sport fish, targeted by both recreational anglers and commercial fishers. Despite this pressure, the population appears stable for now.

QUEEN PARROTFISH *Scarus vetula*

Size: commonly 13 inches (33 centimeters), up to 24 inches (61 centimeters)

Food: algae on coral and rocks, occasionally live coral, sponges, and other animals

Range: coral reefs, rubble, and rocky areas in the western Atlantic, particularly the Caribbean, Florida to northern Brazil; found 3 to 80 feet deep (1 to 24 meters)

Queen Parrotfish are active in the daytime. These unusual fish have powerful jaws and fused teeth, like a parrot beak, that are used to scrape algae off rocks. They also scrape live coral, which they grind into sand with their throat teeth. Then they poop out sand that eventually becomes tropical beaches! By eating fast-growing sponges, parrotfish keep the reefs from being overgrown. Like all parrotfish, Queen Parrotfish change their sexes over their lives. When young adults, they are females and are reddish brown with a white stripe. During the second part of their adult life, they become "supermales," and are blue and green. A supermale may have a harem of several younger females. They are not a threatened species, but their populations have decreased near Jamaica and Haiti due to overfishing.

STONEFISH *Synanceia verrucosa*

Size: up to 16 inches (40 centimeters)

Food: small fish, shrimp, and other crustaceans

Range: coral reef flats or under rocks, ledges, and plants, even tide-pools; widespread from the Indian Ocean to the western Pacific to the Red Sea to East Africa; found 0.5 to 98 feet deep (0 to 30 meters)

Stonefish are members of the scorpionfish group, which are named for their stinging spines. This stonefish species, also commonly called the reef stonefish, is the world's most poisonous fish. They have been known to accidentally kill children and elderly people. There are thirteen needle-sharp back spines, so strong they can pierce a boot sole. Each spine has two sacks that release poison for protection, not for catching food. Stonefish hide their camouflaged bodies in rock or coral crevasses, sometimes partially burying themselves in sand. They wait in ambush to suck in small fish, shrimp, or crabs that pass by. Oddly, they are also aquarium pets. These stonefish are a popular delicacy to eat in Japan, China, and Hong Kong as a cooked fish, and sometimes as sushi, if the spines are carefully removed first.

PACIFIC HAGFISH *Eptatretus stoutii*

Size: 12 to 25 inches (30 to 64 centimeters); females are typically larger than males

Food: any dead, rotting fish or large mammals (whales, seals) that float down from above; or live marine worms, sea stars, shrimp, crabs, or fish in traps

Range: silty and muddy seafloor of the eastern north Pacific; typically found 131 to 350 feet (40 to 100 meters), but also down very deep 500 to 3,000 feet, (160 to 914 meters)

Often called "slime eel," hagfish are not eels but rather a very ancient fish with no bones, no scales, no jaws, no fins, and no eyes, and an ability to absorb nutrients through their skin. As a scavenger, they find their often-dead food by touch and smell, although swarms of hagfish are known to attack and eat fish that are caught in traps. They can grab hold of the food, tie themselves in and out of knots, and make enough leverage to rip off a chunk. Or they enter the prey's mouth or scrape their way into the dead animal, then eat it from the inside. Their loose skin and knots that roll down their body also make it hard for attackers to keep hold of them. Hagfish are eaten in Korea, and their skin (called eel skin or yuppie leather) is made into soft leather wallets, purses, and belts. They provide an important role in the underwater habitat because they clean up dead organisms.

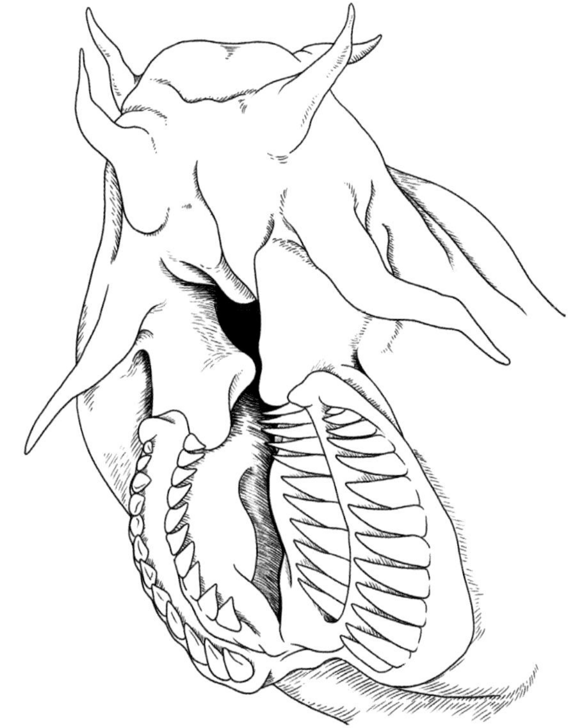

AN OCEAN OF PLASTIC

EEWWW!! That's Really YUCKY!

Plastic sure is convenient, and it doesn't break when dropped. Imagine if shampoo bottles were glass and broke in the shower! We can bring bottled water with us almost anywhere. Take-out meals come with plastic forks and straws that can be tossed in the trash. When we purchase something at a store, we are almost automatically given a plastic bag, except in places where they are banned.

The convenience of plastic comes with a huge price, though. That plastic bag from the grocery store takes ten to twenty years to decompose. Even when disposed of properly, long-lived plastic trash often ends up in streams and rivers that eventually flow to the ocean. Plastic breaks down into tiny colorful pieces that are easy to swallow, either accidently or on purpose, by seabirds, turtles, whales, and so many more innocent animals. Organisms get stuffed with these indigestible pieces of trash and then don't eat enough real food. Many species cannot survive with even a small amount of plastic in their bodies. Plastic marine litter kills as many as 1 million seabirds and 100,000 sea turtles every year; even whales and fish die from eating large or tiny bits of plastic. Plastic bags floating in the sea look a lot like jellyfish and endanger animals who eat jellyfish. Many animals also get tangled in plastic rope, packaging, and old nets. Lost or discarded fishing nets, nicknamed "ghost nets," continue to drift at sea, threatening, catching, or drowning marine fish and mammals.

According to the EPA (US Environmental Protection Agency) our appetite for the convenient items made of plastic has grown over the years. In 1960 we threw away 390 tons (a male polar bear weighs one ton) of plastic; in 2018 we threw away 35,680 tons in the United States alone. One estimate says 33 billion pounds of plastic leaks into the marine environment from land-based sources every year—roughly equivalent to dumping two garbage trucks full of plastic into the oceans every minute.

Much of the plastic waste produced in the United States ends up in landfills or is shipped to poor countries, with only 9 percent of plastic recycled globally. In the process of shipping recycled plastic in large containerships to other countries, 10 percent of that (8.8 million tons per year) ends up in the sea. A humongous mass of plastics twice as big as Texas (over 620,000 square miles) is floating in the middle of the Pacific Ocean. It is nicknamed the Great Pacific Garbage Patch. The trash moves on currents like a conveyor belt and builds up in several places. Overall, it is estimated that 165 tons of debris are circulating in oceans today.

Even balloons cause problems. When full of helium and released, they fly skyward, but when they deflate, they land somewhere. An animal may eat it or get tangled after it deflates. We need to stop pretending that balloons just disappear into the sky and realize that what we are doing is littering. Balloon releases are just another way for plastic to end up in the environment.

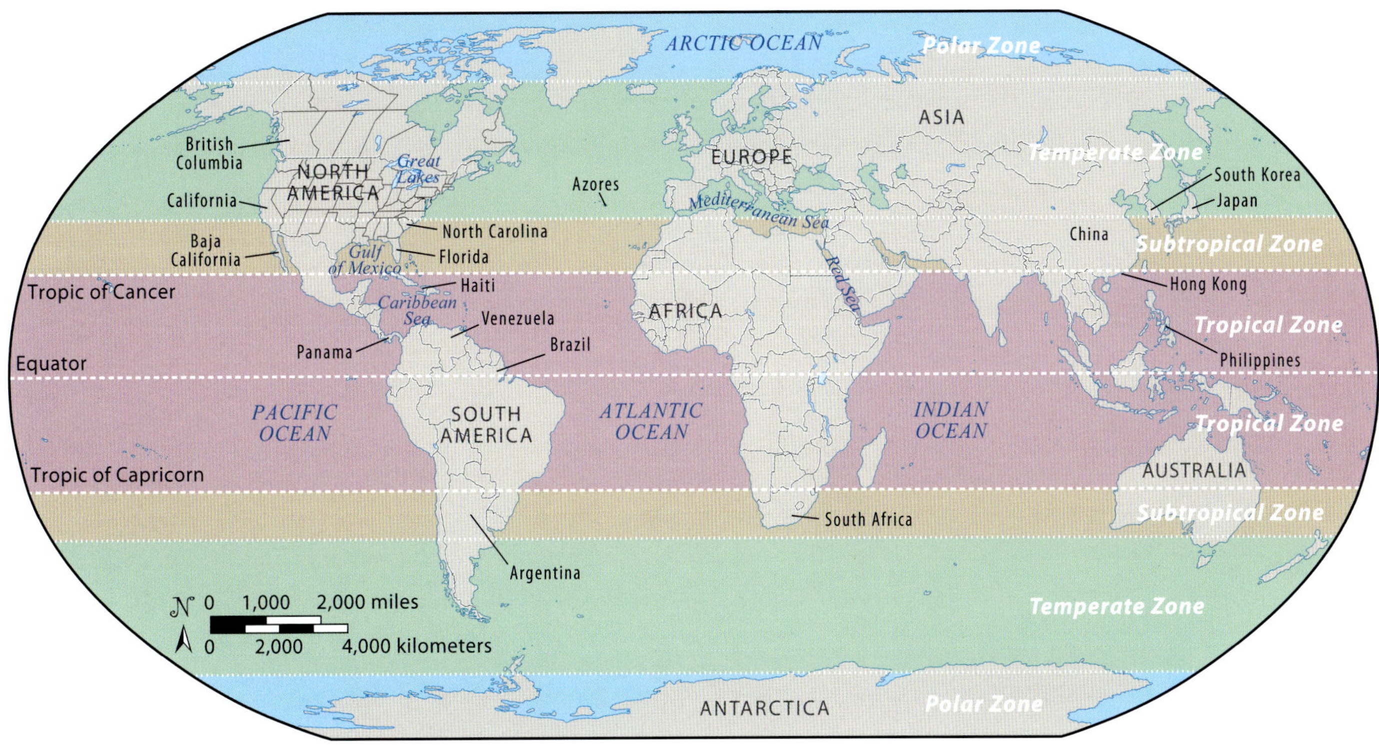

WHAT YOU CAN DO TO DECREASE PLASTIC POLLUTION AND HELP WILDLIFE

Take a survey of all the plastic that you use at home. Pay attention to packaging, bottles, and other throwaway plastic. See if you and your family can begin to reduce these purchases.

- Minimize purchasing single-use plastic bottles and disposable silverware.
- Properly recycle plastic bottles, containers, packaging, and bags.
- Encourage your family to carry cloth bags to all stores.
- Carry a reusable water bottle.
- Use reusable plates, silverware, and glasses at picnics and while traveling.
- Use metal straws instead of plastic ones.
- Talk to your friends, family, and teachers about the reasons all balloon releases should stop.
- Participate in or create a clean-up event and sort what you find into categories to see what and where it comes from.
- Work with school leaders or parents to petition your city to ban plastic bags or start an effective plastics recycling program if your city doesn't have one now.
- Adopt a storm drain in your neighborhood; pick up nearby litter before it goes down the drain.
- Buy items made from recycled plastics such as clothing, furniture, or building materials.

Can you think of other actions? Write to us on our Facebook Page *Nature's Yucky children's books.*

Just for the Halibut

Halibut are a type of flatfish that live on the ocean floor. They are born with one eye on each side of their head. Then as they start to grow up, one eye moves slowly to the other side of its head. *EEWWW!! That's Yucky!*

To help you appreciate this neat behavior, you can make this recipe. In addition to making it for lunch, it's great for *Nature's Yucky!*–themed birthday parties.**

You will need:

One half of an English muffin

A small package of cream cheese to spread on the muffin

Two raisins or black olive slices for eyes

Deli meat or cheese cut into 1-inch circles with a small cookie cutter or a bottle cap

Carrot cut into mouth and fin shapes

Green grapes or lettuce strips for seaweed

**Many grocery stores carry both dairy-free cream cheese and gluten-free bread for those with dietary issues.

GLOSSARY

algae. Plant-like living things, such as seaweed, that grow primarily in water and use sunlight for food and produces oxygen. In the ocean, algae produce nutrients that they pass to coral cells. Then the corals produce waste that the algae eat.

camouflage. A defensive trait that living things use to disguise their appearance, usually to blend in with their surroundings.

cannibalism. The act of an animal eating other animals of the same species.

carnivores. Animals that eat only the meat of other animals.

copepods. Small, shrimp-like crustaceans that are found in most water environments, such as ponds, lakes, and the sea.

coral. A living thing that looks like a plant but is actually an animal that remains in one place and lives in groups in tropical waters.

crustaceans. Any of a large group of mostly water-dwelling creatures that have a hard, jointed shell, such as lobsters, shrimps, and crabs.

gill. An organ used by animals such as fish for getting oxygen from water.

invasive. A living thing, such as a parasite, plant, or animal, that is introduced to an environment and has the potential to cause harm to the environment as it spreads.

kelp. Large, brown seaweeds. Giant kelp grow up to 215 feet and provide important habitat for sea life.

larvae. A form of an animal after birth or hatching that is different from its parents.

mollusks. Animals with a soft body, often enclosed in a shell, such as snails, clams, and octopuses.

nocturnal. Active during the night and resting during the day.

parasite. A living thing living in, with, or on another living thing in order to feed off of it, grow, or multiply and can either directly or indirectly harm the animal it is feeding on.

plankton. A living thing that cannot swim well enough to move against tides and currents and instead drifts with them. There are two main types of plankton: phytoplankton, which are plants, and zooplankton, which are animals. Both are very important to the ocean food chain.

predator. A living thing that kills and consumes other living animals.

prey. An animal hunted by a predator for food.

reef. A chain of rocks or coral, or a ridge of sand at or near the surface of water.

scavenger. A living thing that usually feeds on dead or decaying bodies of other living things.

shellfish. An animal without a backbone that lives in water and has a shell, such as oysters and crabs.

skates. A type of ocean fish that are mainly flat, ray-like, and slow-moving and live on the ocean floor.

RESOURCES

Organizations for More Information or Actions

Ocean Preservation Society, www.opsociety

The Ocean Conservancy, www.oceanconservancy.org

Oceana, https://oceana.org/

National Geographic Kids, https://kids.nationalgeographic.com/pages/topic/ocean-portal, https://kids.nationalgeographic.com/nature/kids-vs-plastic/article/ocean-plastic-by-the-numbers

Smithsonian National Museum of Natural History. Ocean: Find Your Blue portal, https://ocean.si.edu

National Oceanic and Atmospheric Administration (NOAA) fisheries program https://www.fisheries.noaa.gov/welcome

NOAA's Marine Debris program, https://marinedebris.noaa.gov

World Wildlife Fund, www.worldwildlife.org/initiatives/ocean

The Nature Conservancy, www.nature.org/en-us/what-we-do/our-priorities

Healthy Seas, http://healthyseas.org/

Conservation International, www.conservation.org/priorities/doubling-ocean-protection

Encyclopedia of Life, https://eol.org

Fish Base, https://fishbase.se

Top Twelve Aquariums in the United States

Aquarium of the Americas, New Orleans, LA. https://audubonnatureinstitute.org/aquarium

Aquarium of the Pacific, Long Beach (Los Angeles), CA. www.aquariumofpacific.org

Florida Aquarium, Tampa, FL. www.flaquarium.org

Georgia Aquarium, Atlanta, GA. www.georgiaaquarium.org

Monterey Bay Aquarium, Monterey, CA. www.montereybayaquarium.org

National Aquarium, Baltimore, MD. https://aqua.org

New England Aquarium, Boston, MA. www.neaq.org

Oregon Coast Aquarium, Newport, OR. https://aquarium.org

Ripley's Aquarium of the Smokies. Gatlinburg, TN, Myrtle Beach, SC, or Toronto, Canada. www.ripleyaquariums.com

Seattle Aquarium, Seattle, WA. www.seattleaquarium.org

Shedd Aquarium, Chicago, IL. www.sheddaquarium.org

Tennessee Aquarium, Chattanooga, TN. https://tnaqua.org

Books for Adults

Barraclough, Susan. 2007. *Sharks and Other Creatures of the Deep*. New York: Backpack Books.

Beer, Amy-Jane, and Hall, Derek. 2007. *The Illustrated World Encyclopedia of Marine Fish and Sea Creatures*. London: Lorenz Books, Anness Publishing Ltd.

Grant, John, and Jones, Ray. 2006. *Window to the Sea: Behind the Scenes at America's Great Public Aquariums*. Guilford, CT: Insider's Guide, Globe Pequot Press.

Kaplan, Eugene H. 1999. *Peterson Field Guide: A Field Guide to Coral Reef of the Caribbean and Florida*. New York City: Peterson Field Guides, Houghton Mifflin Co.

Smith, C. Lavett. 1997. *National Audubon Society Field Guide to Tropical Marine Fishes of the Caribbean, the Gulf of Mexico, Florida, the Bahamas, and Bermuda*. New York City: Borzoi Book, Alfred A, Knopf, Inc.

Children's Books

Barner, Bob. 2015. *Sea Bones*. San Francisco, CA: Chronicle Books LLC.

Bredeson, Carmen. 2010. *I Like Weird Animals: Leafy Sea Dragons and Other Weird Sea Creatures*. Berkeley Heights, N.J.: Enslow Elementary, Enslow Publishers, Inc.

Brown, Laaren. *Animal Bites: Ocean Animals*. Animal Planet. New York City: Liberty Street, Time Inc Books.

Fiedler, Heidi. 2015. *Up Close Sea Life: A Close-Up Photographic Look Inside Your World*. Lake Forest, CA: Quarto Publishing Group USA Inc.

Hughes, Catherine D. 2013. *First Big Book of the Ocean*. Washington, D.C.: National Geographic Kids, National Geographic Society.

Jackson, Tom. 2014. *The Magic School Bus Presents: Sea Creatures*. New York City: Scholastic Inc.

Prager, Ellen J. 2014. *Sea Slime: It's Eeuwy, Gooey and Under the Sea*. Mt. Pleasant, SC: Sylvan Dell Publishing.

Rhodes, Mary Jo, and Hall, David. 2007. *Survival Secrets of Sea Animals*. New York City: Children's Press, Scholastic Library Publishing.

Rizzo, Joanna. 2016. *Ocean Animals: Who's Who in the Deep*. Washington DC: National Geographic Society.

Shea, Nicole. 2012. *Creepy Sea Creatures (Nature's Creepiest Creatures)*. New York City: Gareth Stevens Publishing.

Team Ocean. 2015. *Really? Ocean*. New York City: Scholastic Inc.

Both Karen (left) and Lee Ann (right) are longtime educators of children and adults. They give presentations on writing children's books. They both love to look for these "yucky" animals when snorkeling and scuba diving in tropical waters.

Landstrom is the retired director of Eastman Nature Center with the Three Rivers Park District in suburban Minneapolis. She has a master's degree in biology, was past president of the Minnesota Association for Environmental Education, served for many years on the board of the Minnesota Naturalists' Association, and is currently a board member of her local National Audubon Society chapter. She continues to lead eco-trips to Costa Rica and to take international trips to see bird, wildlife, and exotic habitats. She enjoys gardening, biking, bird-watching, photography, and giving travelogue presentations. Contact Lee Ann and find more information at https://www.leeannlandstrom.com/.

Shragg is retired from her thirty-five-year career as a naturalist and naturalist manager in the Twin Cities Metro Area. Before that, she was both an elementary teacher and worked with preschool children. Karen now runs MUSEC LLC, (Move Up Stream Environmental Consulting) and writes about her passion—how to save wildlife—and other conservation issues. She holds a bachelor's degree in elementary education, a master's degree in outdoor education, and a doctorate in critical pedagogy. Karen enjoys bird-watching, nature travel, kayaking, and paddle boarding. She can be reached via her website, www.movingupstream.com.

Rachel Rogge studied Art History, Biology, and Science Illustration at Humboldt State University in Northern California, and Science Illustration at UC Santa Cruz. Rachel has been working as an illustrator since 1999, with a focus on science and natural history topics. She currently lives in the Pacific Northwest, where she enjoys watching all manner of wildlife on land and at sea.

The previous *Nature's Yucky!* titles have won the following literary awards: Best Books, Southwest Book Award, the National Association for Interpretation's Media Award, and finalist for the Minnesota Book Award.